BRIDGES OF SEATTLE
IMAGES of America

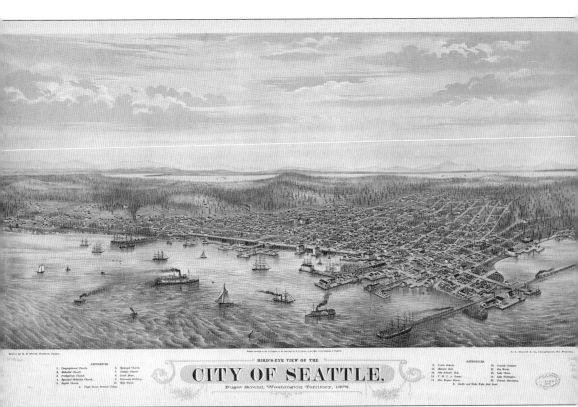

This 1878 aerial perspective map of downtown Seattle shows the surrounding hills and bodies of water that presented considerable challenges to the city as it grew outward from its original center. Reshaping the landscape and innovative bridge design were instrumental to connecting Seattle's neighborhoods. (National Archives.)

ON THE COVER: Workers resurface Seattle's historic Fremont Bridge with open steel-mesh deck grating in this 1936 image. Patented in 1932 by Walter F. Irving, steel-mesh grating improved the safety of formerly timber-decked bridges. Its first use in the United States was on Seattle's University Bridge in 1933. (Seattle Municipal Archives, item 10763.)

IMAGES of America
BRIDGES OF SEATTLE

Maureen R. Elenga

ARCADIA
PUBLISHING

Copyright © 2020 by Maureen R. Elenga
ISBN 978-1-4671-0438-8

Published by Arcadia Publishing
Charleston, South Carolina

Printed in the United States of America

Library of Congress Control Number: 2019943030

For all general information, please contact Arcadia Publishing:
Telephone 843-853-2070
Fax 843-853-0044
E-mail sales@arcadiapublishing.com
For customer service and orders:
Toll-Free 1-888-313-2665

Visit us on the Internet at www.arcadiapublishing.com

To Sofia and Otto, my bridges to the future

CONTENTS

Acknowledgments		6
Introduction		7
1.	The First Steel Bridges	9
2.	The Lake Washington Ship Canal	17
3.	The Ship Canal Bascules	33
4.	Depression-Era Bridge Projects	65
5.	Postwar and Contemporary Bridges	99
Bibliography		127

ACKNOWLEDGMENTS

Unless otherwise noted, all images appear courtesy of Seattle Municipal Archives and will be cited as SMA followed by the item number. I owe a debt of gratitude to Julie Irick, photo archivist at Seattle Municipal Archives, for her help and patience with my frequent requests. My research for this book was aided by the outstanding resource HistoryLink.org, the online encyclopedia of Washington State history. I would like to thank the contributing historians whose work I turned to repeatedly. Among them are Paula Becker, Alyssa Burrows, John Caldbick, Phil Dougherty, Glenn Drosendahl, Linda Holden Givens, Jim Kershner, Greg Lange, Kit Oldham, Jennifer Ott, Alan J. Stein, Cassandra Tate, David B. Williams, and David Wilma. Thank you.

Introduction

Seattle's natural environment is defined by steep hills and large bodies of water, which made topographical alterations and bridge construction essential to overcoming the challenges to growth presented by its landscape. The land on which Seattle was built is hourglass-shaped, with downtown located at its narrowest point, pinched between Elliott Bay to the west and Capitol Hill and Lake Washington to the east. To the north is Lake Union, and to the south, the Duwamish River empties into Elliott Bay, the deep-water inlet of Puget Sound on which Seattle was founded, forming a vast tidal flat.

For thousands of years, Coast Salish indigenous people lived in villages of cedar structures along the edges of Puget Sound and its tributaries and estuaries. They navigated its myriad waterways in dugout canoes as a means of swift transportation between village communities, fishing and hunting areas, and agricultural prairies located along and within the densely forested landscape. Over the centuries, the Puget Sound Salish developed a rich and complex culture that left little imprint on the physical environment.

The arrival of Seattle's first permanent white settlers in the early 1850s brought significant changes to the landscape bordering the tidal flats of Elliott Bay, beginning with the deforestation of land claims for block platting and to initiate Seattle's timber export industry.

The tidal flats to the south and steep hills to the north and east hemmed in the young community on a small area of flat, buildable land. The early settlers who followed established homesites that were widely scattered among the surrounding hills. Although Lakes Washington and Union, their waterways, and the Duwamish River delta separated these settlements from the commercial core, canoeing and rowboating between communities and shipping out of Elliott Bay were sufficient to meet the demands of transporting people, goods, and commerce for the area's first two decades of slow growth.

With periods of steady and sometimes booming growth in the last two decades of the 19th century, the topographical barriers presented significant challenges to expansion and transportation, resulting in massive reengineering of the landscape and, sadly, displacement of most of the indigenous population who remained in the area.

Although Seattle did not gain competitive transcontinental rail service until 1893, railroad speculation in the 1880s sparked the city's first real estate boom. By the end of that decade, population growth had pushed residential areas farther north and east of an expanding downtown district located in the area that is now known as Pioneer Square. Seattle's population grew from 3,533 in 1880 to 42,800 in 1890, necessitating major infrastructure development and improvement. These infrastructure projects were preceded by a massive fire in June 1889 that leveled downtown Seattle and initiated an unprecedented period of construction and population growth that continued until the nationwide financial panic of 1893.

The Klondike Gold Rush of 1897 pulled the city out of the slump of the panic. Seattle's successful effort to promote itself as the location for prospectors to procure supplies and passage to Alaska brought an economic and population boom that was sustained through the first decade of the 20th century, pushing the population from 80,671 in 1900 to 237,194 in 1910.

One of Seattle's first timber trestle bridges was the two-mile-long Grant Street plank road, which spanned the tidal flats between downtown and South Seattle. Because the city was undertaking street regrading and tidal flat–filling projects to ease growth and increase buildable land, bridges were built exclusively of pile and timber trestle with the expectation that more substantial, permanent crossings would be financed upon the completion of the topographical improvements.

Between 1910 and 1920, Seattle's population grew to 315,312. The period coincided with the dawn of the automobile era, ushering in the city's most ambitious era of bridge construction. Seattle's first permanent steel bridges were constructed in 1911, including the Twelfth Avenue South or Dearborn Street Bridge, Washington State's oldest steel arch bridge.

Since Seattle's earliest days, city leaders had endeavored to connect Lake Washington to Puget Sound, but political and financial obstacles prevented progress on a comprehensive project. The construction of the Lake Washington Ship Canal took place from 1901 to 1911, allowing passage from Lake Washington to Salmon Bay, but a wider, deeper canal and locks were needed to facilitate navigable passage in and out of Puget Sound. Federal funding came through in 1910 and enabled the construction of the shipping canal and locks proposed by Hiram Chittenden, head of the Seattle District of the Army Corps of Engineers.

The locks project necessitated the replacement of existing timber trestle bridges with permanent drawbridges, which would allow for unobstructed passage of ships along the eight-mile canal route and accommodate the increasing flow of automobile and streetcar traffic to neighborhoods north of downtown. Four drawbridges were completed on the Lake Washington Ship Canal between 1917 and 1924.

While private and commercial construction slowed to a near halt during the Great Depression, Works Progress Administration (WPA) projects supported ongoing bridge construction and improvements throughout those difficult times and provided the funding that paved the way for Seattle's legacy of engineering innovation in bridge construction. Thanks to these innovations, Seattle is home to some of the world's only permanent floating pontoon bridges, including the world's longest floating bridge.

The bold and ambitious undertakings to reshape Seattle's landscape and connect its communities to the central business district in the early 20th century have left the city with a wealth of beautifully designed bridges that stitch together distance and time while serving outstanding natural beauty to generations of residents and visitors. Bridges are an integral part of the Seattle experience whether one is taking in views of the Cascades to the east and Olympics to the west while crossing the Aurora Bridge or Mount Rainier from the floating bridges that span Lake Washington.

With more than 250 vehicular and pedestrian bridges maintained by the city's department of transportation, Seattle is a bridge enthusiast's delight. The chapters that follow tell the story of some of the city's most historically significant bridges. Many of the bridges included are listed in the National Register of Historic Places and are designated city landmarks.

One
THE FIRST STEEL BRIDGES

Seattle's massive regrading projects took place between 1898 and 1931; the alteration of street levels to ease growth and movement throughout the city also necessitated the construction of some of Seattle's first permanent steel bridges. This 1914 image of a steam shovel at work regrading Sixth Avenue at Marion Street illustrates the dramatic alteration of the environment. (SMA, item 37.)

The Yesler Way Bridge is Seattle's oldest permanent steel roadway bridge. Regrading lowered the street levels on Terrace Street and Fifth Avenue, so the Yesler Way Bridge was built over Fourth Avenue in 1910 to ease the east–west grade at Yesler Way's intersection with Terrace Street. The ornamental capitals of the fascia girder columns and decorative pedestrian railings are shown in this 1962 photograph of the bridge. (SMA, item 69857.)

Improvements on the Yesler Way Bridge were completed in 2017. The work included removal of its interior support columns to create a single-span superstructure and preservation of its historic decorative details. Cars are parked along Terrace Street by the triangular 400 Yesler Building in this 1920 image, with the north side of the Yesler Way Bridge visible below them. (SMA, item 1756.)

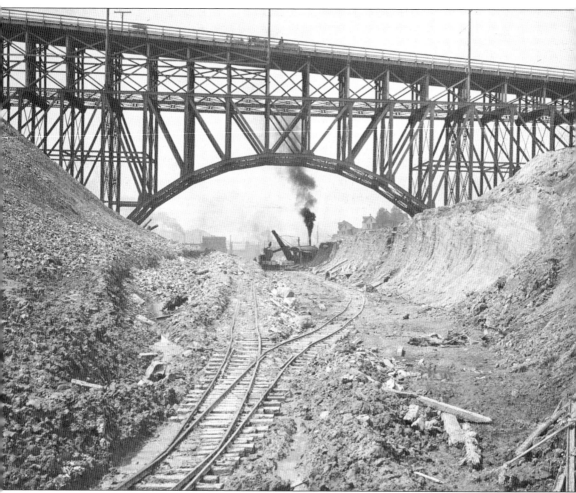

In 1911, one year after the completion of the steel bridge at Yesler Way, the Twelfth Avenue South (or Dearborn Street) Bridge was constructed. The steel arch bridge connects Twelfth Avenue South to Beacon Hill, crossing over Dearborn Street. The area the bridge spans was excavated in a sluicing project, undertaken by Seattle city engineer R.H. Thompson, that removed portions of Beacon Hill in order to connect downtown with the Rainier Valley residential areas to the south. The 171-foot bridge has Pratt-style web trussing and cantilever spans of 94 feet and 96 feet at the south and north ends, respectively. Work on the Dearborn Street regrade is ongoing in this 1912 picture taken shortly after the completion of the Twelfth Avenue South Bridge. (SMA, item 6070.)

In 1917, during heavy spring rains, the southern approach to the Twelfth Avenue South Bridge was destroyed by a mudslide that shifted the bridge 30 inches north. The damage is visible in this image taken prior to the city's repair of the bridge and rebuilding of the timber approach that same year. (SMA, item 5709.)

In 1924, the timber approaches to the Twelfth Avenue South Bridge were replaced with concrete approach spans, and its streetcar lines were removed. The regrade of Beacon Hill to accommodate Dearborn Avenue is complete in this 1937 image, and rail line tracks have been removed in preparation for paving. The Twelfth Avenue South Bridge is Washington's oldest steel arch bridge. (SMA, item 11422.)

The next extant permanent steel bridge constructed in Seattle was not necessitated by regrading projects. It was built to span a deep ravine in Ravenna Park at Twentieth Avenue NE. Ravenna Park is located in a half-mile-long, 115-foot-deep wooded ravine in the Ravenna neighborhood, north of the University District, and is part of a contiguous green space connecting with Cowen Park. The city acquired the parkland in 1911 after the 1907 annexation of Ravenna. The historic park was once home to old-growth timber, as shown in the 1911 image at right. By 1926, the trees had all been removed. In the 1909 image shown below, children wade in Ravenna Creek, which runs through the ravine, as two adults and a baby look on. (Right, SMA, item 30088; below, SMA, item 30112.)

Constructed in 1913 to link the growing residential areas on either side of the park, the Ravenna Park Bridge is the older of Washington's two remaining three-hinged, steel lattice-arch bridges. The bridge is considered one of the finest legacies of Seattle city engineer Arthur Dimock, whose tenure encompassed the city's most active period of new bridge construction. Designer Frank M. Johnson oversaw the design team, which opted for the more costly steel arch design over more economical forms, as they felt it would better integrate into the park setting. The 354-foot-long bridge has a 250-foot double-rib arch that stands 41 feet above the ravine at its apex. Scalloped fascia plates on the underside of the deck echo the form of the double-rib arch. Originally built for vehicles, the bridge has served only pedestrians since 1975. (SMA, item 417.)

Two

THE LAKE WASHINGTON SHIP CANAL

On July 4, 1854, near a lake called Tenas Chuck ("little waters"), Thomas Mercer addressed a gathering of Seattle's residents. Mercer proposed naming the larger lake to the east Lake Washington and changing the name of Tenas Chuck to Lake Union. He chose "union" because he envisioned the lake being a link to a canal connecting Lake Washington to Puget Sound. Lake Union is pictured from its north shore in this 1903 image. (SMA, item 172577.)

Seattle business and civic leaders were eager to make Thomas Mercer's vision of a canal from Lake Washington to Puget Sound a reality; they believed the ease of shipping goods through a calm, protected inland waterway would provide for economic growth and foster industrial development away from the increasingly crowded piers of the downtown waterfront on Elliott Bay, shown here

in 1892. In the decades that followed, as the city experienced growth in its population and its timber, fish, and coal export industries, the need for more waterfront land also grew. However, progress toward this goal would come in fits and starts through private enterprise until a large monetary commitment from the federal government could be secured. (SMA, item 65590.)

David Denny, who owned a lumber mill on the south shore of Lake Union (shown below in 1885), formed the Lake Washington Improvement Company with other area landowners in 1883 to build canals between the two lakes and between Lake Union and Salmon Bay. That same year, the company contracted to dig out the slow, narrow stream that was Lake Union's outlet to Salmon Bay. The flow of water to the small canal was controlled by a wooden dam, lock, and spillway. Upon the completion of the canal in 1885, the company constructed another canal with two timber-pile locks between Lakes Union and Washington that provided passage for small boats and was used as a log chute. Another 35 years would pass before a navigable connection to Puget Sound was completed. (Left, Queen Anne Historical Society; below, Washington State Digital Archives.)

City government efforts to secure federal funding ultimately displaced privately funded ventures. In 1891, the US Army Corps of Engineers renewed consideration of a canal linking Lake Washington to Puget Sound—a project it had abandoned 20 years prior—in earnest. The corps made a decision about the route, as shown above on the 1895 map, and the necessary land was condemned for right-of-way and ceded to the government in 1900. The corps chose the route north through Salmon Bay, shown below in 1900, to Shilshole Bay in favor of a route from Salmon Bay south through Smith Cove. The Smith Cove route had the strategic advantage of more inland protection from possible wartime bombardment and a closer proximity to the downtown waterfront, but the potential cost of construction far exceeded that of the Shilshole route. (Above, Washington State Digital Archives; below, SMA, item 170353.)

MAJ. CHITTENDEN COMING

Engineer Officer Is Selected to Succeed Maj. Millis, in Charge of Seattle River and Harbor District.

WASHINGTON BUREAU
of
THE SEATTLE SUNDAY TIMES.
and
THE PUGET SOUND AMERICAN.
901 Colorado Bldg., Saturday, March 10.

MAJOR HIRAM M. CHITTENDEN, of the corps of engineers, has been selected to take charge of the Seattle river and harbor district, to fill the vacancy caused by the transfer of Major Millis to the Philippines. He is now stationed at Sioux City and will go to Seattle upon being relieved by Major Quinn, who is now in Savannah as a witness to the Green-Gaynor trial.

Major Chittenden was appointed to the military academy from New York in 1880, graduated in 1884 and advanced through the various ranks and was commissioned major in January, 1904. He served as lieutenant colonel of volunteers in the Spanish War.
W. W. JERMANE.

In 1906, Maj. Hiram M. Chittenden came to Seattle to head its district of the US Army Corps of Engineers and got to work planning the canal. He made changes to the 1891 recommendations, which called for locks between Lakes Union and Washington to maintain the depth of Lake Washington. Chittenden's plans called for two masonry locks at the west-end narrows of Salmon Bay with no lock to prevent lowering the level of Lake Washington to that of Lake Union. Lowering Lake Washington would redirect drainage west, away from its natural outlet south through the Black River. Chittenden's plans were approved by the USACE in 1908, but funding had not yet been allocated. The Salmon Bay site is pictured below in 1915. (Left, *Seattle Times*; below, National Archives.)

Excavation for the Montlake Cut between Lake Washington and Lake Union began in 1909, through state and county funding, and was completed in 1910. The corps realized that a longer, wider, and deeper canal was needed. Excavation for the larger canal, pictured above in 1913, began in June 1912 and was completed by late 1913. On December 31, 1913, the cofferdam at the west end of the canal was removed, sending a deluge into the cut. Persistent erosion problems necessitated the building of new cofferdams and draining of the canal so concrete could be poured to stop the erosion in 1914 and again in 1916. (Above, SMA, item 6496; below, Washington State Digital Archives.)

This September 1914 image shows work being done on the Montlake Cut between the construction of the second and third cofferdams. The photograph was taken from what would later be the location of the Montlake Bridge, the fourth and final drawbridge built along the Lake Washington Ship Canal's route. (SMA, item 390.)

On August 26, 1916, the west cofferdam was removed. A few days later, the east cofferdam was removed, and the two lakes were finally joined by the Montlake Cut. Lake Washington eventually lowered nine feet, drying up its natural outlet, the Black River. This was a devastating loss to the Duwamish Indians, who had lived and fished along the river's edge for centuries. (Washington State Digital Archives.)

Federal funding for the canal and locks came through a $2,275,000 allocation in the 1910 River and Harbor Act. Construction on the locks began in August 1911 with the building of a 2,300-foot-long curved cofferdam around the construction site on the north end of Salmon Bay. Crews dredged a temporary channel south of the dam for the passage of boats. The cofferdam was completed in August 1912. The lock pit was then dredged and drained, and a 65-foot-high supply train trestle was built through its center. Two gantry cranes spanned the width of the construction site and rose 75 feet above the supply train. Two movable cars on each of the gantry cranes were capable of hauling five tons each and allowed workers to transport heavy supplies to and from wherever they were needed. (Above, SMA, item 6316; below, SMA, item 6324.)

The corps began work on the Fremont Cut connecting Lake Union to Salmon Bay in early June 1911; it is pictured above with a steam shovel at work in the distance. The cut followed the path of the small canal dug by the Lake Washington Improvement Company in the 1880s. The first cut maintained the natural creek's function as an overflow drain for the lake. The new cut raised its water level to that of the lake and became the lake's conduit to Salmon Bay, which, upon completion of the locks, was transformed from a saltwater inlet to a freshwater harbor. The Fremont Cut is 5,800 feet long and 270 feet wide with a navigable channel down its center that is 100 feet wide and 30 feet deep. (Above, Washington State Digital Archives; below, SMA, item 179383.)

With the lock pit construction infrastructure in place, the work of pouring concrete for the structure of the lock took place from early 1913 through late 1914. As the concrete work neared completion, installation of the lock gates began. This December 1914 image shows the steel framework of the large lock's downstream upper service gate prior to the application of its steel sheathing. (National Archives.)

This June 1915 image shows the upstream-side upper guard gate of the large lock with its steel sheathing in place. In all, nine sets of gates were installed in the locks: three sets for the large lock, two for the small lock, and one set of guard gates at each end of both locks. (National Archives.)

The concrete work in the lock pit and the installation of the gates, pictured above in 1915, were completed by early 1916. In February 1916, the large lock was opened, and crews began work on a cofferdam in the temporary canal at the south end of the locks to enclose the area for construction of the spillway. The 235-foot-wide spillway allows water to overflow from Salmon Bay as needed to keep its level and that of Lakes Union and Washington at 20 to 22 feet above sea level. With the completion of the spillway dam, the gates of the locks were closed, completing Salmon Bay's transformation to a freshwater harbor. After three weeks, enough fresh water had filled Salmon Bay to allow boats to pass though the locks, which officially opened on August 3, 1916. (Both, National Archives.)

On July 4, 1917, 63 years after Thomas Mercer announced his vision for a canal between Lake Washington and Puget Sound, Seattle held a grand-opening celebration for the Lake Washington Ship Canal. The SS *Roosevelt* led a parade of 200 boats to Lake Washington through the ship canal, which was lined with spectators. Hiram Chittenden was too ill to attend the celebration of the locks he planned; he died three months later. (National Archives.)

While the federal government funded and oversaw construction of the ship canal and locks, the city undertook infrastructure projects that were necessary to support the increase in land and water traffic. The projects included road improvements, bridge construction, sewers, walkways, drainage, and street regrading, as shown here on Shilshole Avenue in June 1914. (SMA, item 244.)

Renamed the Hiram M. Chittenden Locks in 1956, the locks facilitated the passage of more than 16,000 vessels in their first year of operation. Today, the locks serve mostly pleasure craft and light industry, with shipping activity concentrated at the Port of Seattle, which was established in 1911. The locks served the ship- and aircraft-building industries during World War II, as illustrated in this 1944 image of a B-29. (National Archives.)

The spillway's role in keeping the lakes and Salmon Bay at a steady, even level made possible the construction of floating bridges across Lake Washington. The first of those bridges, the Lake Washington Pontoon Bridge (later renamed the Lacey V. Murrow Bridge), is shown on this c. 1940 postcard. (National Archives.)

Three
THE SHIP CANAL BASCULES

Federal funding for the construction of the ship canal and locks came with a mandate that Seattle fund and construct permanent steel and masonry bridges along the canal to allow the passage of oceangoing vessels by the time the locks were operational. The city's efforts resulted in the four double-leaf bascule bridges in use today. The first to be completed was the Fremont Bridge, which was finished in 1917 and is pictured here in 1928. (SMA, item 47003.)

The Fremont Bridge location, at the north end of Lake Union, was originally occupied by a rickety, low pile-trestle bridge built in 1890 that spanned the small canal dug in the 1880s by the Lake Washington Improvement Company to connect Lake Union to Salmon Bay. The bridge was built for streetcars to serve the growing communities of Fremont and Ballard to the north. It connected these communities to downtown via a two-mile-long trestle that ran along the west shore of Lake Union. The bridge was also used by pedestrians, wagons, and the occasional automobile. The old Fremont Avenue Bridge was razed in 1911 for widening and dredging of the ship canal. A higher trestle bridge, constructed as a temporary replacement, opened in 1912. Surface planks for the replacement bridge are being laid in this June 1911 image. (SMA, item 130317.)

Prior to the completion of the replacement bridge at Fremont Avenue, another temporary bridge was built five blocks east at Stone Avenue. Business owners in Fremont fought the proposal for the Stone Avenue bridge, concerned that it would divert traffic away from Fremont's business center. Advocates for the Stone Avenue location prevailed, and the bridge, named the Stone Way Bridge, was approved in May 1910. (*Seattle Times*.)

FREMONT SURPRISED BY BRIDGE APPROVAL

Decision of Board of Public Works in Regard to Stone Avenue Structure Opposed by Many in Suburb.

CALL FOR TENDERS TO BE ADVERTISED

Residents of District That Would Be Benefited by Lake Crossing Form Club to Boost Its Cause.

The Stone Way Bridge became the Fremont neighborhood's first high bridge when it opened on May 31, 1911. The timber trestle bridge was 2,700 feet long and 25 feet above the water, with a steel-truss opening for boats. Spanning the northwest corner of Lake Union, the bridge was wide enough to accommodate the ever-increasing streetcar and vehicular traffic between Seattle and its suburbs north of Lake Union. (SMA, item 1376.)

The two temporary spans at Fremont and Stone Avenues effectively served pedestrians, vehicles, and streetcars for the following two years. Any resentment Fremont business owners may have harbored toward the Stone Way Bridge was eliminated on March 13, 1914. Early that afternoon, high water in Lake Union caused the Fremont dam that controlled its level to fail. The usually placid water below the Fremont Bridge became a torrent that washed out its center portion. In the below photograph, observers stand on the banks of the lake as water flows into the canal toward the Fremont Bridge. The breach lowered Lake Union by eight and a half feet in 24 hours, washing out docks, stranding boats, and leaving houseboats hanging precariously above water. The US Army Corps of Engineers repaired the dam, which held for the remainder of the canal project. (Above, SMA, item 575; below, SMA, item 99.)

The Stone Way Bridge, visible in the distance at upper right in the above picture, held during the dam breakage; floodwaters are shown flowing toward Fremont in the foreground. With the Fremont Bridge rendered useless, the Stone Way span took on all of its traffic, providing vital access for Fremont businesses and residents. Crews quickly repaired the damaged bridge, and it reopened to wagons and pedestrians about six weeks later. The Fremont Bridge streetcar lines that had been rerouted to the Stone Way Bridge continued to operate over the Stone Way trestle until the completion of the permanent Fremont Bridge. In the below image, a streetcar approaches the Stone Way Bridge. (Above, SMA, item 100; below, SMA, item 52193.)

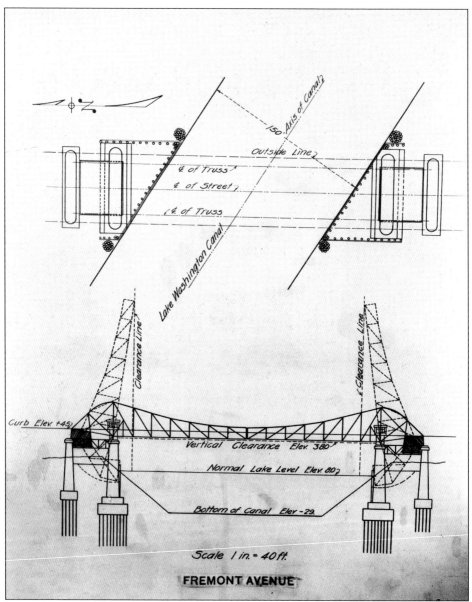

City leaders had to decide what type of replacement would best meet the requirements of the federal government and serve the long-term interests of the city. One month after the Fremont dam failure, city engineer Arthur Dimock presented his recommendation to city council. Dimock spent two years studying various options, including fixed, swing, vertical lift, and bascule bridges, as well as the possibility of tunnels. He determined that tunnels were too costly and problematic to build. The amount of land that would need to be acquired for the long approach to a fixed span 150 feet above the water would be costly; vertical lift bridges would require obtrusive 200-foot towers; swing bridges were ugly and occupied too much land. He recommended Chicago-style double-leaf bascules, modeled after bridges built in the 1890s by the Chicago Public Works Department, for all spans over the canal. Bascules require little land for their approach, and their leaves open to near-vertical position for unlimited clearance, as illustrated in this engineering department plan dated April 24, 1914. (SMA, item 165.)

Bascule comes from the French term for "teeter-totter," a reference to the way the movable spans, or leaves, are lowered and lifted by counterweights made of concrete-filled steel boxes housed underneath them on transverse girders at the back end of the trusses. One-hundred-horsepower motors move the spans independently, setting in motion the counterweights, which rotate around a pin called a trunnion. (National Archives.)

The temporary Fremont Bridge, shown here in March 1915, was closed in August 1915 for construction of its permanent replacement. The cost of construction was estimated at $342,000. The finished bridge would be 502 feet long and 30 feet above the water. As with all bascules on the canal, the 45-foot-wide road deck would accommodate two streetcar tracks, two vehicle lanes, and two sidewalks. (SMA, item 587.)

Throughout the construction of the Fremont Bridge, North End citizens pushed for the Stone Way Bridge to stay in place. Puget Sound Traction, Light & Power, which ran streetcars on the bridge, also fought its removal. In a last-ditch effort, city council adopted a resolution on March 17, 1917, stating "that a permanent bridge across Lake Washington Canal at Stone Avenue is necessary for the traffic and commerce of the people of the North End, and the war department is hereby requested to designate said Stone Avenue as the site for a permanent bridge and to permit the retention of the present wooden bridge until water-borne commerce requires its removal, or for at least two years, provided a wooden draw be installed at once." The effort failed, and the Stone Way Bridge was ordered to be removed. The Stone Way Bridge is visible in the distance, with the north-end pier of the Fremont Bridge rising in the foreground. (SMA, item 1133.)

The city was given until October 1917 to remove the trestle bridge at Stone Avenue. The Fremont Bridge officially opened on June 15, 1917. In the May 1, 1917, image above, workers are laying pavers six weeks prior to the bridge's opening. The completed bridge is shown with both of its movable spans open in the August 1917 image at left. The construction of the next bascule bridge to be completed on the Lake Washington Ship Canal, the Ballard Bridge, did not stir similar controversy within the community it served, as it was replacing an existing bridge one block adjacent to the location of the new one. However, it did play a part in the debate over Ballard's annexation to Seattle. (Above, SMA, item 1311; left, SMA, item 139480.)

Ballard existed as a separate municipality from the time of its incorporation in 1889 until 1907, when it was annexed to Seattle. Citizens of Ballard debated the annexation issue for a few years, first voting against it in 1905. Chief among their concerns was securing a connection to Seattle's freshwater supply from the Cedar River watershed and building a proper sewer infrastructure. The State Supreme Court ruled that Seattle did not have to provide Ballard access to its water supply, pushing the debate further toward annexation. By 1906, although the US Army Corps of Engineers' plan for a canal was two years away from approval, the eventual construction of a canal was a foregone conclusion. Ballard, which operated from the city hall pictured here, was concerned that Seattle would refuse to pay anything but a small amount for the bridge that would have to be constructed. With these infrastructure concerns mounting, Ballard voted in favor of annexation and became part of Seattle on May 29, 1907. (SMA, item 11935.)

The first bridge connecting Seattle to Ballard was a wagon road bridge built in 1889; it served for a couple of decades until it rotted and was replaced with a fixed trestle. The second, built in 1890, was a railroad trestle serving the neighborhood on the north shore of the bay. Both early bridges are shown in this 1903 image looking north across Salmon Bay. (SMA, item 30032.)

The fixed trestles crossing Salmon Bay, shown here at low tide before the bay's transformation to a freshwater harbor, obstructed work on the Lake Washington Ship Canal. By 1909, they had been replaced with a single bridge that allowed for the passage of barges during work on the canal and handled all rail and vehicle traffic between Seattle and Ballard. (SMA, item 833.)

New Drawbridge at Head of Salmon Bay

Needing the federal government's right-of-way clear for work to begin, the War Department issued Seattle a permit in October 1909 to build a temporary drawbridge that was to stay in place until the completion of the higher, permanent Ballard Bridge. The city utilized parts of the old trestles to construct approaches to the temporary bridge, which spanned Salmon Bay between Thirteenth Avenue West and Fourteenth Avenue NW in Ballard. Known at first as the Salmon Bay Drawbridge, it was also called the Fourteenth Avenue NW Bridge. At its center was a Howe-truss swing span that opened perpendicular to the road deck to allow boats to pass. The bridge was certified complete on June 14, 1910. The opening of the Salmon Bay Drawbridge, which would serve the city for eight and a half years, was announced on June 20, 1910, with this *Seattle Daily Times* headline and image. (Seattle Times.)

Being a part of this massive infrastructure project was a source of pride for those involved. In this June 1, 1911, photograph, men who worked on the Lake Washington Ship Canal between Ballard and Fremont pose in front of the North Queen Anne streetcar with a sign reading "Ballard Canal Brigade." All the men have signs reading "Ballard" tucked into their hat straps. (Washington State Historical Society.)

Construction of the Ballard Bridge began on September 1, 1915. As with the Fremont Bridge, the massive piers that supported the movable steel truss spans were constructed of concrete, and the approaches were timber trestles. The roadway was made from blocks of creosoted wood, which protected the wood from rot but was slippery when wet. Work on the pier forms is shown in this 1916 image (SMA, item 1097.)

Work on the ship canal began four years before work started on the Ballard Bridge. This lack of coordination between the City of Seattle and the Army Corps of Engineers resulted in more costs for the city. In February 1916, with work only halfway done on the Ballard Bridge, Lt. Col. James Cavanaugh informed Seattle mayor Hiram Gill that the gates of the locks would be closed that coming July to elevate the water level of Salmon Bay, completing its transformation to a freshwater harbor; to avoid its being inundated, the Fourteenth Avenue NW Bridge would have to be raised. City engineer Arthur Dimock estimated the cost at $5,000. In April, the Seattle City Council passed an ordinance to raise the bridge and appropriated $5,800 for the job. The alterations were completed in time to avoid the temporary bridge being submerged on July 12, when the lock gates were closed and Salmon Bay began to rise. (SMA, item 1362.)

The 295-foot Ballard Bridge opened on December 15, 1917—six months to the day after the opening of the Fremont Bridge, its fellow Chicago-style bascule bridge to the east. In the January 28, 1918, photograph above, citizens celebrate the formal opening of the Municipal Street Railway service on the newly completed Fifteenth Avenue NW Bridge (the street name for the Ballard Bridge). With the Fourteenth Avenue NW Bridge still in place, large ships could not enter the ship canal beyond Salmon Bay. In August 1918, the War Department ordered it removed, and three months later, the city passed an ordinance for it to be done. In the March 13, 1918, image below, a ship passes through the Ballard Bridge from the east side of Salmon Bay; the Fourteenth Avenue NW Bridge is visible in the background. Evidence of the former bridge's location remains today in the gravel parking strip dividing the lanes of Fourteenth Avenue NW where trolley tracks once ran. (Above, SMA, item 12506; below, SMA, item 130310.)

It would be more than a year and a half before the third bascule bridge on the Lake Washington Ship Canal, the University Bridge, was completed. Bonds were put before voters in June 1914 to fund construction of permanent drawbridges at Ballard, Fremont, Sixth Avenue NE, and Montlake, but funding was only approved for the Ballard and Fremont crossings. Debate arose over the crossing location at Sixth Avenue NE, where existing trestles were located. A proposal to fund a bridge at Tenth Avenue NE, a more direct crossing into the University District, passed in March 1915. In this 1911 image, a car drives along Interlaken Boulevard with a view to the University of Washington across Lake Union's eastern extremity, Portage Bay. The Tenth Avenue NE Bridge, later named the University Bridge, would angle across at the narrow northern bend in the lake, which is visible at left across the water from the house in the foreground. Funding for the Montlake Bridge would not be approved by voters until 1924. (SMA, item 29374.)

The existing spans at Sixth Avenue NE, together referred to as the Latona Bridge, were built in 1891 and 1902. The first was constructed by pioneer David Denny for a streetcar line into Brooklyn (later called the University District). By 1902, Denny's streetcar line and his fortune were lost. The first bridge, now owned by the city, was converted for pedestrian and vehicle traffic. The adjacent span was built by the Seattle Streetcar Company. To facilitate work on the canal, the Latona spans were converted to drawbridges in 1916. The city's bridge had a steel-truss center span that swung out horizontally. The railway bridge opened vertically, as shown above. The neighborhoods served by the bridge were still sparse but growing rapidly. In the 1903 image below, land is being cleared for University Boulevard. (Above, SMA, item 12669; below, SMA, item 172579.)

In 1909, Seattle hosted the Alaska-Yukon-Pacific Exposition on the University of Washington campus, shown above lit up at night. The exposition showcased Seattle's Klondike Gold Rush prosperity and its geographical position as the gateway to the Pacific. It served as a sort of coming-out party for the young city as well as a way to develop the University of Washington's new campus, to which it had relocated from downtown in 1895. The Latona Bridge carried nearly four million visitors to the site on the north shore of Portage Bay by foot, rail, horse, and automobile. It also carried an increasing number of commuters to the growing residential developments in the University District. Houses on University Boulevard—now Montlake Boulevard—are shown in the 1912 image below. (Above, SMA, item 1186; below, SMA, item 6128.)

Work on the Tenth Avenue Bridge was set to begin on May 1, 1916, and be completed in 15 months. By August, however, the construction company to which the contract was awarded had defaulted. Work was complicated by the condition of the soil where the south-end pier was being built; it was too soft and would require deep pilings to reach firm clay. (SMA, item 1500.)

The steel for the superstructure was ready by November, but with no substructure for it to be attached to, it languished along with the progress on the bridge. A second contractor was awarded the work to finish the south pier, which was finally completed in 1918. The bridge is shown here six months before completion with both piers and the steel superstructure in place. (SMA, item 1602.)

The 291-foot Tenth Avenue Bridge, officially renamed the University Bridge in June 1919, was completed on June 29, 1919, nearly two years later than scheduled. The final cost of the bridge was $825,275, more than double what the Fremont or Ballard Bridges had cost. Construction complications and the need to recontract the job at wartime prices had driven costs above the original estimate of $303,000. Fortunately, the steel had already been procured before the United States entered the war, or those costs would have been greatly inflated, too. The University Bridge was formally dedicated on July 1, 1919. In the below image, citizens gather for the celebration as Seattle mayor Ole Hanson drives the municipal streetcar on its inaugural run. (Above, SMA, item 1694; below, SMA, item 12660.)

Voters rejected funding to construct the Montlake Bridge—the final and easternmost bascule bridge on the Lake Washington Ship Canal—multiple times over the course of 10 years until finally voting to approve it in 1924. Excavation for the ship canal bisected Montlake Boulevard, so anyone who wanted to cross the 200-foot gap between the south bank of the canal and the north bank at Montlake had to make a five-mile detour west and cross at one of the other bridges. Ironically, the 83-foot-high concrete foundation piers for the bridge had been in place at the Montlake crossing since 1914, before work even started on the Fremont and Ballard Bridges, as they were built during the ship canal construction on the Montlake Cut. In this 1913 image, men stand at the site of the Montlake crossing looking down at the canal excavation. (SMA, item 6494.)

Although there was clearly a need for a bridge at Montlake, voters rejected funding in 1919 and 1921. Class resentment may have been an issue, considering that the bridge would serve the wealthy residents of Montlake and Laurelhurst. In 1922, bridge boosters looked to the class-defying force of college-football fandom to draw support by promoting it as the "Montlake-Stadium Bridge." The first University Stadium had been built in 1920 a few blocks north of the Montlake Cut on the site where Husky Stadium now stands. The next funding vote, held in May 1922, garnered a majority but not the requisite 60 percent. A vote in 1923 approved funding but was deemed unlawful. Finally, with a public-relations push supported by prominent University of Washington football fans, funding was approved—on the eighth attempt—on March 10, 1924. Love of football was the common bond that finally bridged the Montlake gap. In this 1920 image looking east toward Lake Washington, University Stadium is shown under construction. (SMA, item 12833.)

Construction on the 345-foot Montlake Bridge started on July 8, 1924. Although it is a double-leaf trunnion bascule bridge, like its three predecessors on the Lake Washington Ship Canal, its mechanics differ in the mounting of the trunnions—the cylindrical pins around which the counterweight rotates when the bridge leaves are in motion. The Montlake trunnions are mounted on brackets built off of its concrete footing rather than on transverse girders at the back of the trusses of the moveable spans. The Montlake Bridge, situated at the eastern gateway to the University of Washington campus, also stands out among its peers for its beautiful architectural style and detail. Designed to complement the buildings on campus, the bridge towers were executed in the Collegiate Gothic style. In this March 1925 image, the approaches are in place, and work is being done on the two two-story Gothic towers on the northwest and southeast corners of the bridge. (SMA, item 171047.)

The Montlake Bridge was formally dedicated on June 27, 1925. The completion of the span brought much-needed relief from congestion on the Ballard, Fremont, and University Bridges. With a growing population and a rapidly increasing number of automobiles in the city, traffic on the earlier bascules had increased 40 percent between 1923 and 1924. (SMA, item 10728.)

The completion of the Ballard, Fremont, University, and Montlake Bridges put in place the movable spans that still serve north–south traffic today. They facilitated the growth of the northern suburbs and fulfilled Seattle's commitment in the long effort to build the Lake Washington Ship Canal. In this 1950s image, boats travel east from Lake Union to Lake Washington through the Montlake Cut. (Washington State Digital Archives.)

The booming automobile era necessitated improved surface streets. Earth excavated from the Lake Washington Ship Canal and the city's many regrading projects was used to bury the rail trestles, which had carried commuters over the old bridges, in order to make way for paved roads. Westlake Avenue is being paved in this 1920 image. (SMA, item 1644.)

The engineering efforts involved in reshaping early-20th-century Seattle also forever altered the landscape south of downtown. By 1909, Harbor Island, then the world's largest man-made island, was being constructed at the Duwamish River delta, and the river's tidal flats were being filled to create buildable land. In this c. 1900 photograph, the Grant Street plank trestle is shown over the tidelands with West Seattle in the distance. (Washington State Digital Archives.)

The 1909 construction of Harbor Island divided the mouth of the Duwamish River into two channels called the East and West Waterways. That same year, the city formed the Duwamish Waterway Commission to sell bonds to fund rechanneling the river in effort to alleviate flooding and open the area for industrial and commercial use. The river's natural path was a series of oxbows meandering toward Elliott Bay. Work to straighten and dredge the Duwamish River began in 1913. By 1920, twenty million cubic feet of earth had been dredged from the river. The circuitous, 9-mile-long river was transformed into a straight, 50-foot-deep, 5-mile-long waterway capable of handling oceangoing vessels. This 1915 image shows the East Waterway and looks north toward Elliott Bay. The Smith Tower, completed one year before the photograph was taken, is visible poking above the horizon at far right. (SMA, item 739.)

Four

DEPRESSION-ERA BRIDGE PROJECTS

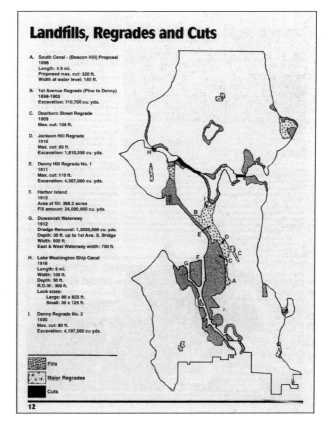

Seattle's major waterway cuts, regrades, and filling projects were completed by the 1930s. The steep hills and flood-prone tidal flats that had been the drawbacks to utilizing Elliott Bay's protected deepwater harbor had been engineered out of existence. This map details these major public works and the years they were completed. (SMA, item 29-001.)

After the first two decades of the 20th century, Seattle's growth and prosperity stalled along with the rest of the country's amid the stock market crash of 1929 and the Great Depression. This 1933 image shows a Hooverville that sprang up south of downtown on reclaimed tideland. (SMA, item 191876.)

The Depression coincided with a period when commuter use of automobiles was surpassing that of streetcars, and horse-drawn vehicles had become obsolete. The automobile era was putting pressure on Seattle's aging spans and increasing vehicle traffic on drawbridges that had to compete with boats in providing passage to a critical point. Improvements and upgrades would need to be made despite the hard financial times. (SMA, item 139491.)

One of the earliest bridge upgrades of the 1930s was at West Garfield Street, where a timber trestle bridge built in 1912 connected the Magnolia peninsula to Interbay over the filled tidelands of Smith Cove. The span was one of a network of trestles built over the rail yards on the lowlands between Queen Anne and Magnolia over the next 20 years. By the time this photograph was taken in 1929, the West Garfield Street Bridge was badly deteriorated. In 1924, one of the three primary routes to and from Magnolia, the Wheeler Street trestle, was burned down by sparks thrown from a train passing below it. A section of the West Garfield Street Bridge collapsed in 1925 under the weight of an overloaded truck. It was repaired sufficiently enough to extend the life of the span for a few more years, but a new crossing to Magnolia would have to be funded. (SMA, item 3391.)

A local improvement district was created, and Magnolia residents were assessed for half of the cost of constructing a new bridge. The remaining half was split between the City of Seattle and the railroad companies. Construction on the new West Garfield Street Bridge is underway in this photograph taken on March 6, 1930. (SMA, item 3871.)

Six months to the day from when the previous photograph was taken, citizens gathered to celebrate the opening of the new West Garfield Street Bridge with a military parade. Magnolia was home to a US Army post, Fort Lawton. Most of the post's property—543 of 703 acres—was given to the city for the creation of Discovery Park in 1973. Fort Lawton officially closed in 2011. (SMA, item 4582.)

Traffic on the Fremont, University, and Montlake Bridges grew so bad that the *Seattle Times* ran a front-page editorial on January 19, 1927, calling for two more bridges across the ship canal. The editorial suggested high fixed bridges at Dexter Avenue north to Stone Way and at Third Avenue West. The cost of the bridge at Stone Way could be shared by the state and federal government, as it would be a necessary link to downtown on the north–south four-lane Pacific Highway (now State Route 99), which was then being constructed. The Third Avenue West location was scrapped, and plans were soon underway for a crossing at Stone Way. The location of the crossing was changed to Aurora Avenue, about 1,000 feet west of Stone Way, because of its shorter distance and firmer foundation. (Above, SMA, item 2857; below, SMA, item 4277.)

The 2,954-foot-long, 70-foot-wide Aurora Bridge is a hybrid concrete-and-steel cantilever deck-arch truss. The center span is a Warren truss suspended between concrete cantilever spans, which are supported by concrete anchor spans. Its concrete piers are shown in this January 1931 image with the Fremont Bridge visible just a few hundred feet to the west. (SMA, item 4716.)

Two men look down from the south-end concrete cantilever span as it awaits connection to the deck truss in this April 1931 image. Because there is no bedrock, the piers are set on log pilings that rest on gravel, sand, and clay under the lake. There are a total of 1,512 logs beneath the north and south piers. (SMA, item 4925.)

Taken from a houseboat community at the southeast edge of Lake Union, this May 1931 photograph shows the Aurora Bridge with its cantilever spans in place before the completion of the center deck-truss span. The open leaves of the Fremont bascule bridge are visible beyond the gap where the center span would eventually be built. (SMA, item 4935.)

The Aurora Bridge marked the passing of an era in both land and sea transportation. Since it was intended to serve an automobile-dependent population, it was the first bridge in Seattle built without streetcar lines. The War Department's 150-foot clearance requirement meant that obsolete tall-masted ships docked at Lake Union had to depart before the completion of the center span or be forever trapped. (Museum of History and Industry [MOHAI].)

The Aurora Bridge, as it is commonly known, was dedicated on February 22, 1932, and officially named the George Washington Memorial Bridge in honor of the bicentennial of the first U.S. president's birth. Thousands gathered on the bridge for its dedication, which included a 21-gun salute and fireboats shooting water into the air. President Hoover unfurled a US flag on the bridge from a telegraph key in Washington, DC. The state and federally funded Aurora Bridge was the first highway bridge built in Seattle and the final link in US Highway 99, which runs from Canada to Mexico. Washington governor Roland Hartley was joined by Mexican consul W.O. Lawton and Vancouver, British Columbia, alderman W.H. Lembke to celebrate the highway connection of nations with the cutting of a Douglas fir log instead of a ribbon. (Above, SMA, item 54372; below, SMA, item 5859.)

In 1932, the same year the Aurora Bridge was dedicated, an effort was underway to rebuild the approaches to the Ballard Bridge. The timber trestle approaches had grown rickety, and its slippery wood surface was the cause of several accidents. But money was tight, and the city decided to rebuild the shaky timber trestle approaches to the University Bridge with funds approved in a 1931 bond measure. A temporary trestle drawbridge was constructed adjacent to the University Bridge to carry pedestrians, automobiles, and streetcars while the approaches were replaced with wider ones constructed of steel and concrete. In the 1932 image at right, the temporary span is shown next to the new approaches under construction. The new, wider concrete approaches are visible in the aerial view below, with the University of Washington campus to the right. (Right, SMA, item 7733; below, SMA, item 77297.)

The wooden decking on the University Bridge was replaced with open steel-mesh grating, as shown here in July 1933; this was the first use of this technology in the United States. The steel grating was a major improvement over timber decking. The bascule leaves were lighter and easier to lift, and the open mesh allowed water to drain, eliminating the slippery conditions of timber decking. The open-mesh decking also resolved the problems that timber decking encountered in high winds, as air can flow through it. The safety advantages of the new decking were dramatic. Previously, the bridge averaged 182 accidents and 6 fatalities every year due to the slippery surface. After the rebuild, no accidents attributed to such conditions have occurred. Although the rebuilding of the Ballard Bridge approaches would have to wait, the city installed open steel-mesh grating on its bascule leaves in 1934, totally eliminating skid accidents on the span. The wood decking on the Fremont Bridge was replaced with steel mesh in 1936. (SMA, item 8071.)

On May 6, 1935, during the height of the Great Depression and with unemployment at 20 percent, Pres. Franklin Roosevelt signed Executive Order 7034, which established the Works Progress Administration (WPA). The WPA, which was renamed the Works Projects Administration in 1939, was part of Roosevelt's New Deal plan to reform financial institutions and restore the economy. The WPA employed out-of-work Americans to carry out infrastructure projects across the country and provided jobs to 8.5 million workers over the eight years of its operation. Among the projects undertaken by the WPA was the construction of 29,000 new bridges and repairs to existing bridges, including several in Seattle. President Roosevelt is pictured above at Warm Springs, Georgia, and First Lady Eleanor Roosevelt is shown below visiting a WPA site in 1936. (Both, National Archives.)

The North Queen Anne Drive Bridge was funded by the WPA and the City of Seattle in 1936 to replace the existing wooden trestle bridge that spanned Wolf Creek Canyon. The new bridge would serve as a connection between the Queen Anne neighborhood and the Aurora Bridge. The bridge site is shown in February 1936 after the removal of the old bridge. (SMA, item 10107.)

Seattle city engineer Clark Eldridge's solution for spanning the 238 feet across the deep ravine was a parabolic two-hinged steel bridge with a 140-foot-high arch. It was the first bridge of its type built in Washington. The dramatic arch is shown in this photograph, with the Aurora Bridge visible in the distance. (SMA, item 10594.)

Seattle city engineer Clark Eldridge's design for the North Queen Anne Drive Bridge was an elegant and cost-effective solution to the challenges presented by its deep, narrow site and by the economy of the time. It was completed by October 1936, after just eight months of construction, at a cost of only $66,119. (SMA, item 191861.)

Schmitz Park in West Seattle is a forest preserve that was donated to the city by Ferdinand and Emma Schmitz in 1912. A timber-truss bridge had carried traffic over Schmitz Boulevard at Admiral Way across a ravine in the park since 1916 but, by 1935, was badly decaying. The old bridge is being prepared for removal in this February 1936 image. (SMA, item 10089.)

City engineer Clark Eldridge again turned to an innovative solution by using hollow concrete cells for the structure of the 175-foot span. Its hollow box cells created a rigid frame and reduced the weight of the bridge, resulting in a span that was 60 percent longer than any previously built using that method. Falsework for the bridge's flattened arch is pictured here in June 1936. (SMA, item 139490.)

The Schmitz Park Bridge was completed in December 1936 at a cost of $134,000—paid with WPA and gas-tax funds. The bridge received national acclaim for its innovative concrete design. A motorist is shown crossing the new bridge shortly after its completion in February 1936. (SMA, item 11412.)

Prior to 1936, two timber trestle bridges—one for pedestrians and one for streetcars—carried Fifteenth Avenue NE over the Ravenna Park ravine and marked the dividing line in the Ravenna–Cowen Park contiguous green space. Together known as the Cowen Park Bridge, these served the Roosevelt and Ravenna neighborhoods north of the University District. In this 1915 photograph, children peer over the footbridge. (SMA, item 30092.)

Clark Eldridge designed an elegant, 358-foot-long open-spandrel reinforced-concrete arch bridge that stands 60 feet above the Ravenna Creek bed. Thousands of area residents gathered under colored floodlights on the evening of June 9, 1937, to celebrate the dedication of the $148,000 Cowen Park Bridge, which was funded by the WPA and state gas taxes. (SMA, item 191795.)

A notable feature of the Cowen Park Bridge is the 12-foot-tall Art Deco light standards adorning the bridge deck. Set back as they ascend to their peak, the lights resemble the skyscrapers of their era. Together, the North Queen Anne Drive, Schmitz Park, and Cowen Park Bridges demonstrate Clark Eldridge's diverse design talents. (SMA, item 191797.)

Funds to rebuild the rickety timber trestle approaches to the Ballard Bridge finally came together in 1937, when the Seattle City Council passed an ordinance authorizing the work. Forty-eight percent of the $800,000 was funded by the WPA, with the rest paid for by Seattle's share of the state gas tax. The bridge is pictured in 1938 with the wooden approaches still in place. (SMA, item 12201.)

Construction on the Ballard Bridge forced the closure of the bridge for 18 months, rerouting traffic that was crossing the west end of the ship canal to the Fremont and Aurora Bridges. A worker operating a bulldozer breaks up asphalt paving on the bridge approaches in this June 1939 photograph. (SMA, item 39022.)

The new Ballard Bridge approaches were constructed of steel and reinforced concrete. In this February 1940 image, steel workers are putting transverse beams in place over deck girders on the half-mile-long southern approach. The steel girders rest on reinforced-concrete piers. The bridge improvements were completed in May 1940. (SMA, item 39456.)

In this 1941 image, cars and a "trackless trolley" drive south over the Ballard Bridge. Seattle replaced streetcars with trackless trollies—buses that run on electric power from overhead wires—that same year. Seattle was the first city in the United States to widely adopt a trackless trolley system, which is still in use today. (SMA, item 39687.)

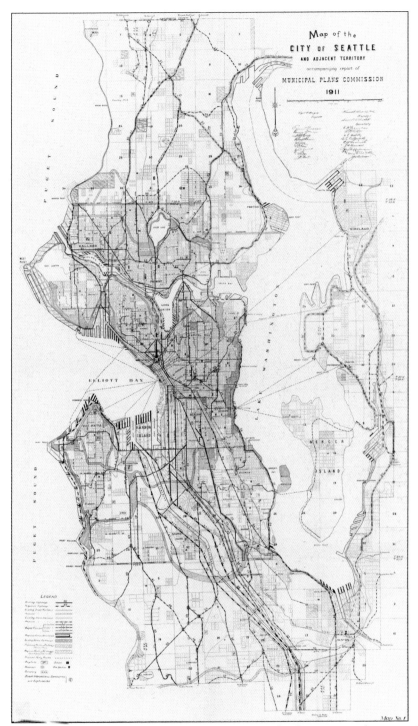

The 1912 Bogue Plan—a comprehensive long-range civic and infrastructure plan created for the Seattle Municipal Plans Commission by consulting civil engineer Virgil Bogue—presented the first serious proposal for crossing Lake Washington via the Yarrow Bay railway tunnel. The plan was approved by the commission but rejected by voters. (SMA, item 609.)

Ferry service between Seattle and Lake Washington's eastern shore began in 1890 and remained the only means of making the crossing for the next 50 years. Mercer Island and the east side of the lake were sparsely populated. The town of Kirkland was the commercial center of the area, and Bellevue was a rural farming community with some shops on its main street. The lakeshore was occupied by a few high-end homes but mostly by rustic vacation cabins. In 1923, a bridge was built between Mercer Island and Bellevue over the east channel of the lake. This was the only means for island residents to drive directly to Seattle by way of a 26-mile trip around the south end of the lake. Ferry landings at Medina (above) and Bellevue (below) are shown in these 1915 images. (Above, SMA, item 226; below, SMA, item 660.)

Lake Washington plunges 214 feet at its deepest point, with 100 feet of soft mud below the water, making a bridge with piers on footings problematic and costly. Proposals to bridge the lake were rejected as poor investments, as the eastside population was too small to justify the cost. For Seattleites, the lakefront was primarily recreational, and few had reason to cross it. Above, a man enjoys the view in a 1913 photograph of Seward Park; below, sunbathers enjoy a public beach in 1930. The first proposal for a floating pontoon bridge came from engineer Homer Hadley in 1921. Hadley worked for a concrete barge manufacturer during World War I and keenly understood the floatation possibilities of concrete. However, Hadley's proposal to construct the world's longest floating bridge garnered no interest from the city or private investors. (Above, SMA, item 29524; below, SMA, item 29786.)

The idea of bridging the lake resurfaced in 1931, when city council held hearings debating four proposed bridges, including Homer Hadley's. The debate continued until a decision was forced thanks to a 1938 WPA offer to fund 40 percent of construction costs; the offer stood through the end of the year. Hadley pitched his idea to State Department of Highways director Lacey V. Murrow. Murrow came out of this meeting sold on the idea and assembled a team of Washington's prominent bridge engineers, who are pictured above. The proposed route from Seattle via tunnels through Mount Baker Ridge created controversy and slowed the process so much that the WPA threatened to withdraw funding unless the city backed the proposed plan. The matter was finally settled, and ground-breaking ceremonies were held on December 29, 1938, two days before the WPA deadline. (Above, SMA, item 73321; below, SMA, item 29569.)

LAKE WASHINGTON BRIDGE PROJECT. E. TUNNEL PLAZA AT 35. AVE.

Construction on the 6.5-mile project started on January 1, 1939. The project included boring two adjacent tunnels through Mount Baker Ridge and constructing west- and eastside fixed approaches, transition spans between the approaches, and the 6,620-foot floating span with a sliding draw pontoon. The pontoon and transition spans are connected to Mercer Island by a 960-foot concrete-and-steel arch bridge. The bridge also required highway construction on Mercer Island, making it an eventual link in the I-90 interstate highway. The pontoons were constructed at graving docks on Harbor Island and guided by tugs through the Government Locks and Lake Washington Ship Canal to the jobsite; Graving Dock 2 is shown in this May 1939 image. The pontoons were connected end-to-end, forming a rigid box girder supporting a 45-foot roadway flanked by 4-foot sidewalks. (Washington State Digital Archives.)

The Lake Washington Floating Bridge project employed 3,000 people during its 18 months of construction. The total $8.85 million cost was divided between the WPA, which contributed $3.8 million, and the State Toll Bridge Authority (STBA). The STBA financed its portion through a bond measure to be repaid through tolling bridge drivers. The west side approach is pictured under construction in December 1939. (SMA, item 73281.)

Mount Baker Ridge is mostly clay, which necessitated a horseshoe-shaped tunnel; its east portal is pictured here in the 1990s. Twin tunnels separate directional lanes of traffic. The tunnels, which were completed in 1940, are each 1,466 feet long with 24-foot roadways and 23 feet of vertical clearance at the apex of the arch. They were the largest-diameter twin soft-earth tunnels in the world when they were completed. (Washington State Digital Archives.)

The east portal of the Mount Baker Ridge tunnel welcomes drivers to Seattle with precast concrete panels featuring stylized Northwest Coastal indigenous motifs designed by James Fitzgerald and Lloyd Lovegren. The center panel reads: "City of Seattle Portal of the North Pacific." Modeling of the precast panels is shown in this 1940 image. (SMA, item 73315.)

The Lake Washington Floating Bridge officially opened on July 2, 1940. Dignitaries and 2,000 invited guests gathered to celebrate the completion of what was then the largest floating structure in the world and the first pontoon bridge built of reinforced concrete. The bridge cut travel time between Mercer Island and Seattle from one hour to seven minutes. (SMA, item 45602.)

The floating bridge became an attraction in its own right. It was so popular that the tolls collected on it paid back the STBA bonds by 1949, which led to the end of tolling 19 years ahead of schedule. The rapid payback was due in no small part to the explosive postwar residential development of eastside communities facilitated by construction of the bridge. The formerly rural and agricultural areas of Mercer Island and Bellevue were transformed into suburban neighborhoods with thriving commercial centers. Between 1940 and 1950, the population and property values on Mercer Island increased over 300 percent. Kirkland's shipyards boomed during wartime but closed in 1946, ending industry on the eastside and concentrating the bulk of postwar development in Bellevue and Mercer Island due to their proximity to the floating bridge. By 1948, the Kirkland Chamber of Commerce was calling for a second bridge farther north to serve its community, but that effort would take another 15 years. A driver is being interviewed during a floating bridge toll survey in this July 1946 image. (SMA, item 40605.)

The Lake Washington Floating Bridge was rededicated as the Lacey V. Murrow Memorial Bridge in 1967 after the death of Murrow, the former State Department of Highways director. A second span, immediately adjacent to the Lacey V. Murrow Bridge, was constructed in 1989 and officially rededicated as the Homer M. Hadley Memorial Bridge in 1993 to honor the man who, in 1921, first envisioned a floating bridge across the lake. In 1990, the pontoon portion of the Lacey V. Morrow Bridge was being refurbished when a storm sent water into hatches that had been left open by construction crews. The original floating section of the bridge sank, pontoon by pontoon, to the bottom of the lake on November 25, 1990. The reconstruction of the pontoon bridge was completed in 1993; the steel arch approaches, shown here in 1960, are its only original features. (SMA, item 63667.)

Five

POSTWAR AND CONTEMPORARY BRIDGES

The postwar years were a time of optimism and innovation in the United States, and Seattle was no exception. Northgate Center, one of the first suburban shopping malls in the United States, opened in 1950. Boeing launched the first successful passenger jet, the 707, in 1954; and in 1962, Seattle held the futuristic Century 21 Exhibition, or Seattle World's Fair, pictured here. Improvement and innovation in bridge design continued as well. (SMA, item 73130.)

Among the first permanent steel bridges constructed in 1911 was the Oxbow Bridge over the Duwamish River on First Avenue South at South Michigan Street. The name referred to the location of its crossing at a meander in the river. The oxbow vanished when the Duwamish Waterway was dug. The low-level swing bridge was shifted and altered in 1916 to span the new channel (as shown in the above photograph, which was taken that year) and became known as the First Avenue South Bridge. The strong current and organisms in the brackish water were constant sources of trouble for the pilings, requiring ongoing repairs and replacement of the machinery in 1929. By the 1950s, industrial growth and increased traffic necessitated a new bridge, and an innovative design was developed to meet the challenges of the site. (Above, SMA, item 1002; below, SMA, item 1005.)

CUT-AWAY PERSPECTIVE
FIRST AVE. SOUTH BASCULE BRIDGE

Another precedent-setting engineering solution was achieved with the new First Avenue South Bridge. The bridge is a double-leaf trunnion bascule type, like most Seattle drawbridges, but its bascule spans are supported by semi-floating cellular reinforced-concrete piers connected by two underwater reinforced-concrete struts, as shown in this cutaway perspective drawing. The piers maintain 85 percent buoyancy by allowing lower-level ports to flood during high tide and drain as the tide recedes, preventing the structure from sinking into the channel's soft mud. The concept of the piers is credited to project engineer Bruce V. Christy, and they were designed by the Seattle Engineering Department. It is the only floating pier bridge in the world. As with many postwar bridges, the First Avenue South Bridge became part of the state's highway system and carries Highway 99. (SMA, item 44737.)

The First Avenue South Bridge, shown here under construction (with the old bridge, still in use, in the background), was built 23 feet higher than the 1911 span to decrease the number of openings. The $6.5 million project was completed in two years, and the bridge was dedicated on September 22, 1956. (SMA, item 52644.)

This 1955 aerial photograph of the First Avenue South Bridge shows its proximity to the Duwamish River's confluence with Elliott Bay. The bridge had the grim statistic of the most traffic fatalities of any bridge in Seattle from the time it opened until a second, adjacent span was built in 1997, splitting the directions of traffic. The two bridges' bascule spans are among the longest in the country. (SMA, item 130316.)

The West Garfield Street Bridge, built between Interbay and Magnolia, underwent extensive repairs in 1953. Its east end, over Fifteenth Avenue West, was fitted with an elaborate grade-separation system in 1957 and 1958 that was designed by Homer Hadley, who had conceived of the floating concrete pontoon bridge across Lake Washington 36 years earlier. The design incorporated one of the earliest uses of prestressed concrete girders in Seattle. In the March 1958 image above, cars on West Garfield Street head toward the westward approach to the bridge, with the elevated approach over Fifteenth Avenue West under construction. In the June 1958 image below, a car passes over the newly completed elevated approaches toward the Magnolia peninsula with Queen Anne Hill visible in the background. (Above, SMA, item 56948; below, SMA, item 57889.)

104

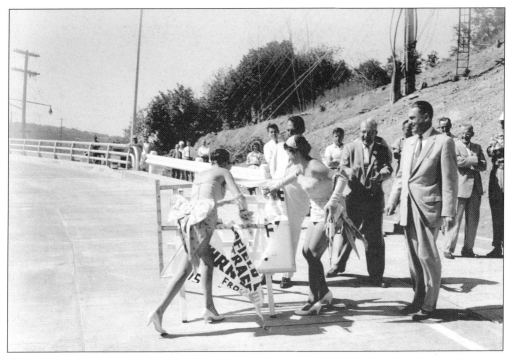

The opening celebration and ribbon cutting for the new West Garfield Street Bridge overpass was held on July 17, 1958. The bridge was officially renamed the Magnolia Bridge in 1960. Extensive repair was required after the 2001 Nisqually earthquake damaged half of the bridge's original concrete lateral bracing. The bridge will be rebuilt when funding is secured. (SMA, item 58009.)

The 1950s brought mounting congestion to Seattle. Its two north–south arterials, State Route 99 and Marginal Way, were proving insufficient to handle the increase in automobiles. Relief came with the 1956 Federal Aid Highway Act, which allocated federal gas-tax funds to pay for 90 percent of an interstate highway network across the country. The highway through Seattle would include a sixth spanning of the ship canal. (SMA, item 43388.)

Plans were soon underway for the Seattle Freeway, which would become part of the Interstate 5 corridor. A 23-mile stretch was cut through the center of Seattle. The Seattle Engineering Department drawing at left illustrates the freeway's path through the heart of downtown. In 1957, the state established an office to acquire property in the right-of-way of the freeway. Most of the 4,500 parcels of land were occupied by homes, apartments, and businesses. The existing buildings were auctioned and relocated or demolished. The process of clearing the right-of-way started at both ends of the future Ship Canal Bridge—the Eastlake neighborhood and University District—and moved along the freeway's path in either direction. Several houses were moved from Lake Union by barge to locations around Puget Sound. A home is being prepared for relocation in the 1959 image below. (Left, SMA, item 61048; below, SMA, item 61730.)

The $16 million Ship Canal Bridge was built just west of the University Bridge, roughly at the location of the former Latona Bridge. Construction of the bridge began with pouring the concrete piers, as shown in this October 1959 image. Like the Aurora Bridge to the west, the Ship Canal Bridge was built high above navigation and required no movable spans. (SMA, item 62947.)

Work on the Ship Canal Bridge's steel superstructure began in May 1960 and continued for eight months. Progress on the superstructure is shown in this September 1960 image, with the University Bridge below the superstructure. Construction of the Ship Canal Bridge was completed by early fall of 1961, more than a year before the completion of the freeway to which it would connect. (SMA, item 63755.)

The Lake Washington Ship Canal Bridge is a steel Warren-truss double-deck bridge. The bridge's six truss sections support a 2,294-foot roadway with eight lanes on the upper deck and four reversible lanes on the lower deck. It is the largest structure of its kind in the Pacific Northwest. The completed overwater span is pictured with Gasworks Park in the foreground. (SMA, item 57750.)

Another link in the network of postwar highway construction came with the long-awaited second floating bridge across Lake Washington. The first floating bridge, pictured here in 1959, brought explosive suburban development to eastside communities. It had reached its capacity of 20,000 cars per day by 1950 and was carrying more than 50,000 per day by 1960. (SMA, item 19859.)

The state legislature approved construction of a second bridge across Lake Washington in 1953. Director of Highways William Bugge announced that six locations were being considered for the bridge. Progress stalled as debate over the location continued throughout the decade. The focus finally narrowed to two routes: one through Montlake to Evergreen Point and one farther north from Sand Point to Kirkland. The US Navy had a base at Sand Point and was opposed to the location. State highway officials and Gov. Albert Rosellini advocated for the Montlake–to–Evergreen Point route, which would connect to the new Seattle Freeway at a point closer to downtown. In 1959, the debate was settled, and bids from contractors were accepted to build at the Montlake–Evergreen Point location. A $30 million bond paid for the bridge was to be recouped by 35¢ tolls. On-site construction began in 1961. Progress on the Montlake span at the west side of the lake is shown in this March 1962 aerial image. (SMA, item 71032.)

The Evergreen Point Floating Bridge, which opened in 1963, was completed over two years at a final cost of $25 million. The bridge had a 7,578-foot-long floating section made up of 35 pontoons held in place by 62 77-ton reinforced-concrete anchors. The roadway accommodated four lanes of traffic. It featured a lift-draw span near the center of its floating section, providing 200 feet of clearance for shipping. Elevated steel truss spans with fixed piers on either end provided clearance for the passage of smaller boats. The new bridge was built four miles north of the first floating bridge and was part of the 5.8-mile road construction project linking the north–south highways on either side of the lake: Interstate 5 on the west and Interstate 405 on the east. The 1.4-mile-long bridge was the largest floating bridge in the world until its replacement was completed in 2016. (Left, SMA, item 29923; below, SMA, item 190659.)

The Evergreen Point Floating Bridge officially opened with a celebration on August 28, 1963. Gov. Albert Rosellini, pictured here, cut the ribbon at the ceremony. The bridge was an overwhelming success. Planners had expected tolling (to repay the $30 million bond) to remain in place until the year 2000. The span was rededicated as the Gov. Albert D. Rosellini Bridge in 1988 but is commonly known as the 520 Bridge for the stretch of highway it carries. The growth it spurred in the communities of Kirkland, Redmond, and Bellevue resulted in a far higher volume of traffic than was estimated, and the bond was repaid by 1979. However, this rapid growth also had the effect of negating the congestion relief that the bridge was intended to provide. These commuting headaches ultimately led to the 1989 construction of the Homer S. Hadley Bridge adjacent to the first floating bridge. (MOHAI.)

The Gov. Albert D. Rosellini Bridge served the expected life of 50 years for which it was built, but by the end of that period, in 2013, it was showing signs of wear and was not up to modern earthquake standards. The bridge was planned with the expectation of accommodating 15,000 vehicles per day, but by 2010, an average of 100,000 were crossing each day. The demands on the bridge necessitated a replacement, and in 2011, tolling was reintroduced on the bridge to pay for a new one. The new bridge was constructed immediately north of the old one; its proximity is shown in this 2015 image. Construction began in 2012, and the completed bridge opened in April 2016. At a total length of 7,710 feet, the new bridge exceeded the length of the old one by 130 feet. The $4.5 billion funding for the bridge came from state gas taxes, federal highway funds, and electronic tolling. (National Archives.)

West Seattle is the largest of Seattle's neighborhoods. It is located on a peninsula west of downtown and separated from the mainland by Elliott Bay and the Duwamish Waterway. The first white settlers landed at what is now Alki Point, where the northern end of the peninsula juts into Puget Sound. After a wet winter, they abandoned the site for the current location of downtown. Early industrial development in West Seattle was centered on its southeast waterfront. Development on the rest of the peninsula picked up with the first real estate boom in the 1880s. Transportation between West Seattle and Seattle was limited to steamship and, later, ferry service until 1890, when the Northern Pacific Railway built a trestle bridge across the Duwamish. This image from the era shows West Seattle in the distance with trestles reaching across the unfilled tidal flats. Spokane Street would become the site of several bridges linking West Seattle, which incorporated as a city in 1902, to neighboring Seattle. (SMA, item 63014.)

West Seattle was motivated to incorporate out of frustration with a real estate improvement company's failure to provide the transportation systems, reliable water supply, and electric light it had promised. West Seattle built its own streetcar system in 1904, but the line could not extend past the city limits. It sold the line to the Seattle Railway Company, which operated Seattle's lines. The streetcar brought a boom in real estate development when it began service in 1907, but the city had made little progress toward getting a reliable water source or electricity. After Seattle annexed the small communities of Youngstown, Alki, and Spring Hill, which shared the peninsula with West Seattle, the vote for annexation of West Seattle passed on June 29, 1907. West Seattle was the largest of the six towns annexed by Seattle that year. Water and electricity followed, as did a new swing bridge built at Spokane Street to replace a temporary 1902 span. Here, a worker is pictured at the Seattle Electric Company's West Seattle substation in 1909. (SMA, item 171376.)

By the time of its annexation in 1907, West Seattle's northern and western shorelines had become popular summer destinations. The new Spokane Street swing bridge provided timely access to West Seattle's attractions. Luna Park, built on pilings over the tidal flats at the mouth of the Duwamish, opened two days before the annexation vote. The glimmering amusement park operated until 1913; after it closed, it was almost entirely dismantled. Its natatorium remained available to swimmers until it was destroyed by fire in 1931. Luna Park's attractions are shown above in 1910. People are pictured below enjoying bathing and canoeing at Alki Point in 1911. Streetcars shared the Spokane Street Bridge with wagons, pedestrians, and a growing number of cars. It was a low-level bridge that required many openings. Demand soon arose for additional crossings to ease congestion. (Above, SMA, item 64761; below, SMA, item 37996.)

A higher swing bridge was built at Spokane Street in 1911. To save on infrastructure costs, the bridge carried West Seattle's water supply from Seattle. This conveyance method required the mains to be uncoupled each time the bridge opened, which cut off the community's water supply until the bridge swung back to a closed position. This considerable nuisance was endured during a period in which the span experienced the largest increase in traffic of all the bridges in Seattle. A 1917 traffic count showed a daily increase in vehicles from 12,400 in 1914 to 21,676 in 1917. A campaign for a high fixed bridge over the waterway was launched in 1916, but the city instead built a second swing bridge, intended to be temporary, in 1918. The 300-foot Howe-truss swing span of the old Spokane Street Bridge is pictured in 1923. (SMA, item 178819.)

Planning for a permanent replacement was underway within a year. City engineer Arthur Dimock cited a 169 percent increase in traffic over the bridges between 1915 and 1919. Direct ferry service between downtown and West Seattle ended in 1921, putting the burden of commuter traffic entirely on the wooden bridges. Rather than the high-level fixed span that West Seattle residents had lobbied for in 1916, city council authorized a steel-and-concrete bascule bridge at West Spokane Street. The bridge was completed in 1924, with a second, adjacent bridge to be built in 1930. In the 1928 image above, a pile of truss members from the 1911 bridge shows advanced decay; in the 1928 image below, the new bascule bridge is in place with the old swing bridge under it. (Above, SMA, item 178825; below, SMA, item 2907.)

A separate bridge for streetcars had been planned by the municipal railway, but the city purchased the streetcar system in 1919, and it was heavily in debt, so the city engineer included tracks in the design of the new bascules with timber trestle approaches. A streetcar is making its way onto the bridge from its west-end approach in this 1928 photograph. (SMA, item 178823.)

The second span was built to the immediate north in 1930. Eastbound and westbound traffic was divided between the two bridges. The bascules served for 50 years, until calls for a high bridge were renewed in the early 1970s amid traffic woes. The city authorized construction in 1972, but it was delayed for years. West Seattle considered secession in 1978, when Mayor Charles Royer suggested fixing up the bascules. (SMA, item 5110.)

The controversy over the Spokane Street Bridge ended at 2:38 a.m. on June 11, 1978, when the freighter *Antonio Chavez*, laden with 20,000 tons of gypsum, crashed into the north span, leaving it in the up position and damaged beyond repair. Negligence on the part of 80-year-old pilot Rolf Naslund was blamed for the collision. Because the bridge could not be fixed, the replacement project qualified for funding from the federal Office of Special Bridge Replacement. Seattle City Council member Jeanette Williams lobbied Congress for the new bridge, and with the help of Sen. Warren G. Magnuson, $110 million in federal funding was secured to build a high bridge over the Duwamish Waterway to West Seattle. Funds from state and local government made up the rest of the cost of around $190 million. (SMA, item 73168.)

The south span of the 1924 Spokane Street Bridge was to stay in place to provide access to Harbor Island and port facilities that would be bypassed by the high bridge. A future replacement for the lower bridge was part of the funding package for the high bridge. Construction on the high bridge began in 1980. It is pictured above rising over the surviving bascule bridge and below nearing completion in 1983. The 2,607-foot-long cantilevered segmental bridge provides 140 feet of vertical clearance. The new high bridge at Spokane Street opened on July 14, 1984. Commonly called the West Seattle Bridge, the span was renamed the Jeanette Williams Memorial Bridge in honor of the city council member who worked for years to get it built. (Above, SMA, item 130330; below, SMA, item 130331.)

The drawbridge that replaced the 1924 bascule bridge marked another milestone in Washington's legacy of innovation in bridge engineering and design. Completed in 1991, the Spokane Street Bridge is the only concrete swing span in the world. It is also the world's heaviest moving structure. Two swing span sections, each 480 feet long and weighing 7,200 tons, are hydraulically operated. The movable spans float on hydraulic oil in steel barrels within the piers, which are located on either shore. The bridge meets with the channel at an oblique angle, requiring each leaf to rotate 45 degrees rather than the 90-degree turn of typical swing bridges. The Spokane Street Bridge received several awards, including a Design for Transportation Award from the National Endowment for the Arts and the 1992 Outstanding Engineering Achievement Award of the American Society of Civil Engineers. It is pictured here with its massive swing spans open to allow passage of a tug moving a barge loaded with shipping containers to the Port of Seattle. (SMA, item 130333.)

The graceful complementary relationship between the design of the West Seattle and Spokane Street Bridges is captured in this recent image. Together, they represent Seattle's past and ongoing efforts to stitch together its communities separated by waterways that present considerable challenges to keeping vehicular and maritime traffic on the move in one of the nation's fastest-growing cities. The Puget Sound region has a long history of ingenuity. As the birthplace of Boeing, Microsoft, and Amazon, its spirit of innovation has changed the world. Seattle's groundbreaking engineering solutions in bridge design are enduring examples of that spirit, which will no doubt continue to thrive on the challenges that come with the transportation demands of the future. (National Archives.)

Seattle's bridges have been a source of civic pride since the first permanent spans were constructed. That pride is reflected in this 1940s postcard, which displays images of the region's natural beauty and feats of ingenuity in its public works projects. (National Archives.)

BIBLIOGRAPHY

Dorpat, Paul, and Genevieve McCoy. *Building Washington: A History of Washington State Public Works.* Seattle: Tartu Publications, 1998.

Holstein, Craig, and Richard Hobbs. *Spanning Washington: Historic Highway Bridges of the Evergreen State.* Pullman: Washington State University Press, 2005.

Kreisman, Lawrence. *Made to Last: Historic Preservation in Seattle and King County.* Seattle: Historic Seattle Preservation Foundation, 1999.

Phelps, Myra L. *Public Works in Seattle: A Narrative History.* Seattle: Kingsport Press, 1978.

Williams, David B., Jennifer Ott, and the staff of HistoryLink. *Waterway: The Story of Seattle's Locks and Ship Canal.* Seattle: University of Washington Press, 2017.

Woog, Adam. *Sexless Oysters and Self-Tipping Hats: 100 Years of Innovation in the Pacific Northwest.* Seattle: Sasquatch Books, 1991.

Discover Thousands of Local History Books
Featuring Millions of Vintage Images

Arcadia Publishing, the leading local history publisher in the United States, is committed to making history accessible and meaningful through publishing books that celebrate and preserve the heritage of America's people and places.

Find more books like this at
www.arcadiapublishing.com

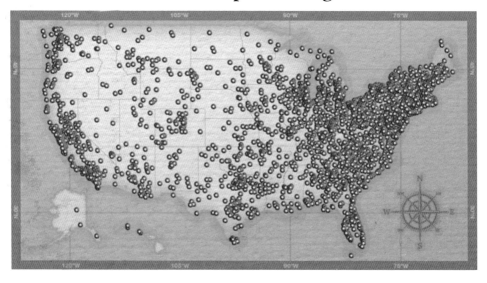

Search for your hometown history, your old stomping grounds, and even your favorite sports team.

Consistent with our mission to preserve history on a local level, this book was printed in South Carolina on American-made paper and manufactured entirely in the United States. Products carrying the accredited Forest Stewardship Council (FSC) label are printed on 100 percent FSC-certified paper.